Copyright © 2021 Clifford Freeman

thespaceglobal.org

All rights reserved.

Edited by Clifford Freeman

Cover art by Clifford Freeman

Layout by Clifford Freeman

Interior graphics by Clifford Freeman

E-Book design by Clifford Freeman

Cover art photograph by Erkan Utu

CONTENTS

Table of Contents

Executive Thinking 4

Decide what you want to talk about 11

    Do People Want it? 11

    How Will You Keep Things Spicy? 14

Name Your Pod! 18

    Search Podcast Name in Google Trends 18

    Search Engine Optimization (SEO) 21

Podcast Design and Branding! 23

    Podcast Colors 23

        Adobe Color Trends 23

        Adobe Create Color Wheel 26

        Podcast Schedule 27

        Podcast Guests 28

The Technology 30

    Software To Facilitate Guest Appearances 30

    Podcast Editing (Mac + Windows) 34

        Editing with a MacBook 35

        Editing with a PC (Windows + Mac) 36

    Hardware For Podcast Production 38

Setting up your Accounts 40

- Make separate podcast email 40
- Social Media Accounts 41
- Where will your Podcast stream? 42
  - Anchor 42
- Setting Up Distribution 44
  - Recording First Episode (the trailer) 44
  - Distributing Podcast 47
  - Amazon Music 51
- Intake Forms and Templates 52

How will you get the word out? 53
- Social Media 53
  - Inner Circle 53
  - Forced Engagement 55
  - Organic Traffic 56
- Paid and Unpaid Advertisements 57
  - Paid Advertisements 58
  - Unpaid Advertisements 58
- Email Marketing 59

Closing 62

About The Author 63

Works Cited 65

Executive Thinking

So I'm going to jump right into it because obviously, you've made it past the point of debating whether or not you need to do this. It's in your destiny. And you have stepped into it.

In The Executive Thinking chapter of How to Start podcasting While Very Busy, I've outlined five major areas that you mostly have to THINK about as you prepare to establish your podcast. Notice how I emphasize "THINK" because, for the next three days, I'm asking you to work these five major items out IN YOUR HEAD alone or with trusted thought partners before moving onto the rest of the book. So, read only the Executive Thinking chapter...give yourself three days to think through this information...then come back to the book for the HOWTO. Ok? Promise? Alright. Let's do this.

Number 1: Decide WHAT you want to talk about. This might sound like a no-brainer but think about this statement carefully:

"Podcasts are literally ALL TALK!"

You need to make sure that the thing you are about to share with the world is something you do not mind talking about, and matter of fact, WANT and CAN talk about in length! Otherwise, you will get bored, and your podcast will become one of those pods that go abandoned after a good 3 episode run, which leads me to my second point about deciding what to talk about. Make sure, especially if you want to do this for a while, that your podcast topic is rich in information! If your podcast is about the color blue, there is only so much spin that you can add before you begin talking about the same stuff over and over or simply do not have anything more to say. Further along in the book, we will discuss some essential keys to picking a topic that is a gift that keeps on giving.

Number 2: The most fun and most trivial: picking a name! I speak about this in length further along in the book because in the age of information and technology, naming something the right thing can be the difference between being seen or hidden. There are many tools out there that can help you indicate if people on the internet will organically search your podcast. In parallel to your name formation process, you also need to start thinking about the branding. In the case of a podcast, that takes three general forms. What will your podcast look like, what will it sound like, and how will it make people feel? This is a very delicate balance, and dance, you have to think real hard about, which should not be taken for granted. We will go deep on utilizing tools that can help you optimize and validate your branding ideas.

Number 3: Setting up your accounts. If you want to run a well-organized podcast, you need to have an organized system. And it turns out that all you need for this organized system to run smoothly is the proper accounts that complement each other to reach your goals. Now, what am I talking about when I say accounts? So you need a communication system for your podcast (email), a form system that keeps track of guests who may come on your podcast (intake forms), the social media platforms you want to use to facilitate your pods' offerings (Twitter, Facebook, etc.), and finally, the most important, the platform your podcast will be streamed on. There are dozens of options that one can pick for all of these categories, and if you are new to the digital content creation world, we will walk through the configuration that will set you up for success.

Number 3.5: Do not think about this one. Just do it now. Make a separate email account. Gmail by Google if you can. Gmail has a multitude of free products that will assist you perfectly with your podcast journey. You want to make sure all of your podcast information is in one place.

Visit:
https://accounts.google.com/signup

Number 4: You need to think about how you want to get the word out about your podcast! There are four primary ways to do this, and considering podcasts are online, 3 out of the four are strategies implemented online. First, first-hand word of mouth/forced engagement. This model is simply directing people you know to listen to your podcast. That one is self-explanatory. I share some pointed guidance on doing that, so you're not coming off as a charity case or a strong-armed robber. The three other major strategies are running ads, email marketing, and strategic social media placement. I speak in length about these three, and depending on what kind of podcast you want to run, I reason with you about what option might be best for you.

Number 5: Finally, the technology! Now, this is a highly contested topic. Since I am a technology enthusiast, people always hit me up asking about what kind of equipment they should use for the project they are trying to get off the ground. For podcasts, I always let them know that it depends on the goals of the podcaster! With the rapid development of technologies like the iPhone, the world of possibility opens up for so many! Especially since podcasts are indeed a multimedia project that utilizes text, audio, imagery, and sometimes video, the iPhone's modern features can put on mighty podcast productions. All of the features are accessible to Android users as well. But like everything else, the almighty iPhone has its limits too. If you want to host multiple guests, or want superior audio quality, then the mobile devices become limited. Depending on your production goals, you want to make sure you have the correct equipment at your disposal. We will go deep on this too.

I remember when I was at the stage you are in now—bundled with more energy than the sun to start this journey. Between you and I, you are the more fortunate one because I didn't have this book to guide me through every step (I'm jealous and proud of you). So please take all of what you have read so far and THINK! And as you think, just remember... you got this; the battle is already won. See you in three days.

This page has been left blank intentionally. When you finish your 3-day brainstorm, please move to the next page.

Decide what you want to talk about

Typically, people don't start podcasts just to be starting podcasts. Usually, the impetus is that the podcaster has something they want to share; some invaluable information that adds a new perspective to understand something in a new way. But sometimes it is none of that. Sometimes you might be a member of a dynamic group of friends, and when you all get together, magic happens. Whatever the case, when you are deciding what to talk about, consider this checklist.

Do People Want it?

Answering this question is very hard to do; as the principal investigator of this question for your podcast, you need to understand that. To help narrow your answer to this question, I give you the skill of research! Utilizing Google Trends!

Google Trends is an absolute powerhouse for making conjectures about what people need (or are looking for online). Google Trends is easy to use, and it is free! Whenever someone types into the Google Search bar on google.com or asks one of their Google-connected smart devices something, Google keeps a receipt. Those receipts are then used to generate what we see in Google Trends. Every minute, Google receives 3.8 million inquiries worldwide (*Google Search Statistics - Internet Live Stats*, 2020). What is to be understood about this number is that Google does not provide answers to these inquiries, per-say, they simply curate the most relevant and viewed weblinks people use and click most often (except if it is a paid advertisement... those always go to the top). But what Google is capturing, though, is the desires of the people utilizing the system. We go to Google every day... all day asking random questions. When we ask the question, we bounce around a few links until we find the info we need. We get information, and Google captures your cyber footprint. This is why Google Trends is so essential. If you wanted to validate if people are looking for specific information, Google Trends could let us know if people are researching the topic.

Google Trends Example: If you have an idea to start a podcast that teaches people "How to Start a Winery," the prompts to ask in Google Trends are something like 'Winery,' 'start a Winery' or 'Winery facts' the keywords and phrases you search for in Google Trends will let you know what people type into Google to find information about. Here are some screenshots of what Google Trends look like for the key terms listed.

Figure 1

According to Google, numbers represent search interest relative to the highest point on the chart for the given region and time. A value of 100 is the peak popularity for the term. A value of 50 means that the keywords are half as popular. A score of 0 means there was not enough data for these keywords (*Google Trends*, 2020).

In Figure 1, the example shows the comparison between 'start winery' and 'winery facts' between December 1st,

2019 and December 2020 in the United States. If these facts were relative to all 50 states, this would be a relatively popular pair of search terms. But when we dig deeper into Google Trends, there is a more telling story.

Figure 2

We observe that 100% of the receipts Google has collected about these two search terms have come from California. So, it is fair to say that in California, the search term winery is booming... and if you are a New Yorker looking to start a podcast to share insights on wineries, not many people in New York will be paying much attention.

How Will You Keep Things Spicy?

After you validate people are looking for the information you will produce, developing a plan to keep the info coming will be your next task. To share how to do that exactly, I want to share the story of why I started my podcast, PhD Tips.

When I was accepted into my Ph.D. program at Boston University in 2020, the first thought that came to my mind was, "What in the world did I get myself into!?" I had no idea or blueprint to follow, as I am the first in my family to complete college and graduate school. When the excitement settled a little, and it came closer to when school would start, my next step was to set up my year one fall semester class folders. In doing so, I realized for my entire higher education career, I organized my learning by year, semester, and class name from freshman year in college until now in my Ph.D. program. I set up an entire portfolio of information that was, by in large, information for me to use as reference points along the way. The podcast topic came to me. I thought to myself, as I stumble and fumble my way through my Ph.D. program, I will record podcasts to share key learning points, milestones, and general tips for others who may land in my shoes one day. And along the way, I will invite current Ph.D. students, Post Docs, and Ph.Ds in the field to share some of their very own insights and tips on "how to get through a Ph.D. program successfully."

The secret here is that the podcast was going to be about the activities of my life. As long as I woke up every morning eager to learn, I would always have content. If you can find something in your life that is rich in information and something people want to obtain or know more about, you favor that it is a good topic to cover. In my case, Ph.D. facts, information, and completion tips were something people inquired about all the time. A pure contribution to the podcast world. Connected to this point, I found two winning solutions you should keep in mind when thinking about podcast content creation. All while the topic at hand is popularly inquired: 1) if you can find numerous experts to frame the perspective of your podcast topic in unique ways and 2) if the topic has zero experts and you have a unique perspective to pioneer for the world; the topic is rich for content creation. For me, having a podcast like PhD Tips, where I can share the stage with a countless number of experts, is a blessing. It is an ingenious strategy to find a podcast topic that has experts because the expertise they bring adds value to your podcast. My theory about successful podcast planning and guest appearances is this: "When you invite professional people to a room to talk about their work, you cannot fail." In many ways, podcasts are a lot like writing your dissertation in doctoral programs. You want to pay homage to experts by inviting them into your space while simultaneously developing original thinking contributing to the field. In the podcast world, you must capitalize on these blessings.

Name Your Pod!

What is a podcast without a name?! Usually, you want to select a name that has something to do with the topic of your podcast. If your podcast is about The Evolution of Humankind, a suitable name for this podcast is something like The Evolution, or The Human Experience, or Evolution. The name of the podcast should give the listener a good sense of what the podcast is about. In addition, I would recommend that you check to see how your podcast name is represented in Google Trends for Search Engine Optimization purposes. Let's go deeper into what I mean.

Search Podcast Name in Google Trends

Now that you are well versed in Google Trends, use it to your advantage! For this example, we will use the name Human Evolution as the podcast name.

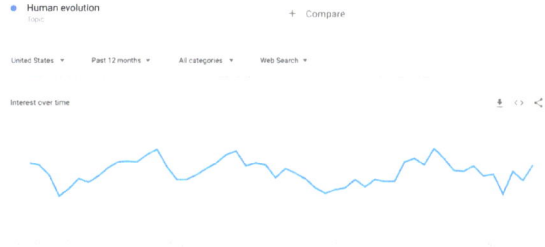

Figure 3

Like Figure 1, Figure 3 displays the search results for Human Evolution in the United States from December 2019 to December 2020. For the most part, we observe that Human Evolution is widely searched.

Figure 4

When we look deeper into the interest by subregion in Figure 4, we see that people all over the country are searching Human Evolution, and Vermont is the state that explores the term the most.

According to Google, interest by region is calculated by the term which was most popular during the specified time frame. Values are calculated on a scale from 0 to 100, where 100 is the location with the most popularity as a fraction of total searches in that location, a value of 50 indicates a location which is half as popular. A value of 0 indicates a location where there was not enough data for this term.

Note: A higher value means a higher proportion of all queries, not a higher absolute query count. So, a tiny country where 80% of the queries are for "bananas" will get twice the score of a giant country where only 40% of the queries are for "bananas."

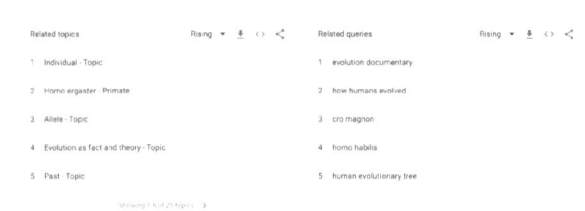

Figure 5

The final figure in this section gives us more information about related search topics. If this screen is populated with as much detail as you see here, it is fair to say that the name you have picked is a very lively topic. Google Trend results like these are the kinds of names you want to choose for your podcast because of the power of organic search and Search Engine Optimization (SEO).

Search Engine Optimization (SEO)

Search Engine Optimization or SEO is a critical topic you must pay attention to. I will not go into significant detail in this book about SEO, as the deep end of SEO is not needed for starting your podcast. What you need to keep in mind when you think SEO is Key Words. Think of SEO this way: When you type in 'Government' in google search while in the United States, what do you suppose will be at the top of the search? Before you check in your Google search results, make a hypothesis. At the time of writing this sentence, I am in Worcester, Massachusetts, 2020, so my best guess is that Donald Trump, Joe Biden, or the United States Federal government dominate that keyword.

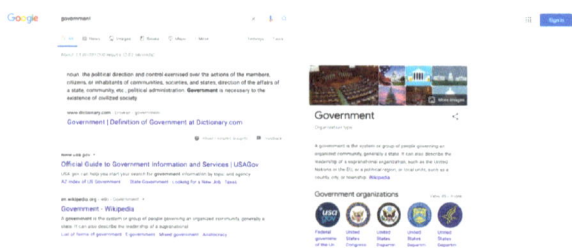

Figure 6

So, it appears my guess was partly correct. The reason I knew the United States government would be high on the Google rankings is that the FED is the most popular and utilized webpage pertaining to the keyword Government in my location. The tricky thing about getting your podcast ranked high in Google search, or any other search engine, is that you compete with whatever is already there. And if the keywords you are looking to rank has no results yet, it takes a few months before you start to rank. As long as people are utilizing your resources and visiting your podcast, then your keywords will begin to rank higher on the charts.

Podcast Design and Branding!

Next up is your Podcast Design and Branding! Like I mentioned in the Executive Thinking section, Podcast Design and Branding takes into account how your podcast will look, sound, and feel. These three features of your podcast ought to be taken care of so that you gain the trust of your listeners.

Podcast Colors

When thinking about your Podcast colors, think creatively! Play a little. Grab a notebook and some markers, and see what colors go well together. If you do not have any notebooks or markers, like me, use technology to help! There are three powerhouse design tools via Adobe Color that you can use, for free, to bring about inspiration and direction for your color strategy.

Adobe Color Trends

Visit:
https://color.adobe.com/trends

Anytime I start a new creative endeavor, and I do not quite know what colors I want to use, I utilize this tool to get different perspectives. The main contribution of Adobe Color Trends is that communities of designers' that crowdsource color palettes that represent "Trends" in specific fields.

Figure 7

If your podcast was about Architecture, as seen in Figure 7, you might want to find the current color "Trends" within Architecture and pick the one you like. You do not have to pick from this library of colors, but I would totally recommend it if you do not already have a particular look you are already going for. Also, because fields tend to converge on the things they know and their own history, so if the field is saying, "we use the colors white and black", you should consider the insight because people already associate the field with that look.

24

Adobe Color Explore page

Visit:

https://color.adobe.com/explore

This page is relatively similar to the Adobe Color Trends page, but it is less categorized by fields. It is more like the explore page on Instagram or Pinterest as a whole. You simply scroll, and scroll....and scroll and scroll until you see something you like.

Figure 8

Alternately, in your case, you may already know what you want to see. If your podcast is going to be about "Outer Space" and you want to know the color palettes associated with it, all you need to do is type it in.

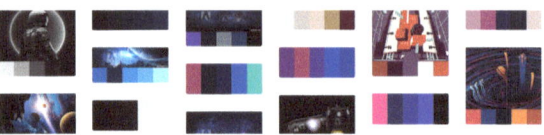

Figure 9

Adobe Create Color Wheel

Visit:

https://color.adobe.com/create/color-wheel

This tool is an extremely powerful design tool that enables you to compare and contrast color palettes of your own. As you noticed in Figure 8, you have the opportunity to save the palettes for later use. What's needed from the pattern to implement into the Color Wheel are the #HEX codes of the color. When you have the #HEX codes to your colors, you then can manipulate your palette to your liking in the color wheel. When you click a particular color pattern, you will see a window like this pop up for you to extract the #HEX codes.

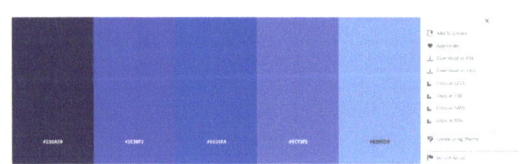

Figure 10

26

Your voice. What will you sound like? Are you going to bring the soft iconic radio voice that calms and soothes listeners, or will you be the electric podcast host that aims to motivate and send jolts of energy to your listeners? How often will you release content? Will it be weekly? Monthly? Daily?! What also needs to be considered is what the host configuration will be? There are more than 850k podcasts out there, and the structures of them are very diverse. Many are solo Podcasts, but a growing number of podcasts are joint operations that have two or more podcast hosts. And what will be your strategy and decision about having a guest? Will you vet them? Will they be friends and family? Something in-between? Let's dive in.

Podcast Schedule

The traditional Podcast is usually a solo show, where the host drops scheduled information on a regular basis one day per week. That is a standard model, which is a logical and efficient model to take up when first starting. As you continue to release content regularly and grow your listenership, people will put you in their weekly routine and expect to hear from you. But some podcasts are completely random when they release new content, and that is ok, too. My podcast, PhD Tips, is sporadic, and it works perfectly that way. The podcast has a lot of recruiting, scheduling, and preparing guests before they jump on the podcast, which leads us into the next topic.

Podcast Guests

As mentioned in the Executive Thinking chapter:

My theory about successful Podcast planning and guest appearances is this: "When you invite professional people to a room to talk about their work, you cannot fail." In many ways, inviting guests to podcasts is a lot like dissertation writing in doctoral programs. You want to pay homage to experts by inviting them into your space while simultaneously developing original thinking that contributes to the field. In the podcast world, you must capitalize on these blessings.

Inviting guests on your podcast is a pivotal opportunity. You want to find guests who are absolute shining stars which adds a perspective you do not already provide. You also have to think of your podcast that you invite guest speakers to as you think of your profile picture on Instagram or Facebook. It should be able to greet and attract anyone who looks and can represent you 24/7 completely. For the sake of the quality of your podcast, you want to invite professional guests, and if given an audience to talk to, they do not need much direction from you to deliver the content the audience is expecting to receive. If you get in the habit of inviting high quality guests to your podcast who contribute different perspectives that are valid and thought provoking, people will see your space as authentic and worth the listen.

The Technology

Technology in this book is referred to as the hardware and software tools associated with recording you and your podcast guest. We will go over some of the equipment you need to get and the software that would speed up your time recording and publishing Podcasts.

Software To Facilitate Guest Appearances

The software I use to convene guests on my podcast is Zencastr. If you plan to convene guests, the number 1 platform to use to capture audio is Zencastr. Before we move forward, please watch this video that explains Zencastr.

Watch video here:
https://vimeo.com/190619561

Ok, so let's do this. First, let's sign up for our Free Zencastr account.

Visit:

https://zencastr.com/pricing

NOTE: You should preference Google Chrome or Firefox as the browser when using Zencastr.

Once you are all set up, let's go over how to create an episode. When you Log-in, your interface should generally look like this.

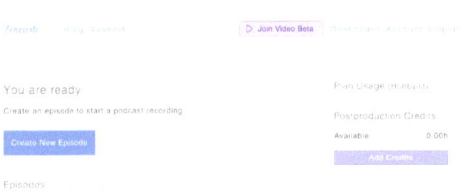

Figure 11

Simply press the big Create a New Episode button to create the episode. Once you type the name of the Podcast episode name, you will be ported right into the session.

The work area for Zencastr should generally look like this.

Figure 12

The interface is drop-dead easy to use. Everything you need is right here on this screen. Let's go over each option.

Start Recording: this button will initiate the recording of your episode. Do not be alarmed; if you press record on accident, it is not live streaming anywhere. The beautiful thing about Zencastr is that every guest you invite, including yourself, gets an individual audio track. Individual tracks are beneficial in post-production because if someone was speaking over someone or speaking too low, you could edit just their individual audio track. If you do this via Zoom, Skype, or some other telecommunication conferencing platform, you get only one audio track with everyone on one file. When your podcast is over, all you need to do is press Stop Recording.

Invite: The invite button is used to send out access to the "episode room" where you interact with guests. There are two options for this. The first option is the option to input the emails and names of the people you are inviting. Zencasr will send them an invite to the "episode room" which they will be directed to join the podcast room.

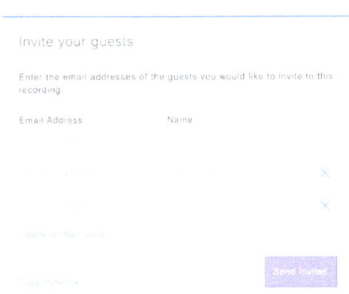

Figure 13

The next option to send an invite is by copying the invite link, as seen in Figure 13. I personally use this option instead of email because I simply insert the link in the calendar invite sent out to guests. When you take this route, it eliminates one extra step for the guest and you to do. In the age of information and technology, too many links and emails get overwhelming.

Timeline Footnote: The timeline footnote in Zencastr is used similarly to how highlighters are used when reading books. When you click the button, Zencastr will take a timestamp. You use this tool when you hear something important and want to use the point in time later. This tool is handy if you do not have any paper and pencil to take notes of important parts of the interview.

Hand Button: The hand tool is a button that illuminates as an indicator to all people on the podcast that you want to say something. This is a good feature for the podcast host to use as a regulator for who has "the mic".

Microphone: The microphone button mutes your microphone. The host of the podcast has the ability to mute anyone in the podcast. Guests only have access to their microphones.

Podcast Editing (Mac + Windows)

The most tedious part of running your podcast will be editing your podcast audio files. If you elect to run a podcast that live streams...then you do not have to worry about editing anything! But in this section, I will share some resources with you depending on if you have a Mac or Windows laptop.

Editing with a MacBook

Any MacBook laptop you have from apple comes with GarageBand. If you do not have the free application on your computer, make sure to download it.

GarageBand Download:
https://www.apple.com/mac/garageband/

GarageBand is an easy-to-use drag and drop, a straightforward, robust, and efficient tool for podcast production. I found an extremely knowledgeable GarageBand Guru on YouTube you should check out to get rolling on GarageBand. Ben Leavitt, is the owner of a popular podcast titled Project Passion, and he shares his Podcast workflow strategy for editing, recording, and exporting your podcast using GarageBand.

Project Passion Podcast:

https://apple.co/2MuaAg3

Ben Leavitt's workflow strategy includes:

1. Folder/file organization
2. How to set audio tracks
3. How to import audio
4. How to cut and merge audio snippets
5. How to record audio

Ben shares a ton of value for users who are just getting started in GarageBand. I would extremely recommend that you take the time to watch his videos and begin to tinker with your own audio files!

Watch how to edit Podcast in GarageBand:

https://youtu.be/n38qu0c-3bs

Editing with a PC (Windows + Mac)

Windows computers do not come preinstalled with any native audio editing software. The most popular free software for Windows users, by far, is an application named Audacity! If you take a liking to Audacity and switch between Windows and MacBook computers, Audacity can be downloaded for both computer types!

Audacity Download:
https://www.audacityteam.org/download

Audacity is a drag and drop, easy-to-use, robust, and efficient audio editing software. I found a very knowledgeable YouTube Channel named Buzzsprout who is passionate about helping independent podcasters find their voice and connect with their listeners.

In Buzzsprout's module, you will learn the following:
1. Getting set up to record
2. How to record in Audacity
3. Record multiple microphones at the same time
4. How to add audio tracks
5. How to edit in Audacity
6. The Selection Tool
7. The Time Shift Tool

8. The Envelope Tool
9. Episode Structure
10. Export podcast from Audacity

Watch how to edit Podcasts on Audacity: https://youtu.be/Zw9nkEHQ5B8

Hardware For Podcast Production

This section keeps in mind that there are two kinds of podcasters. 1) Podcasters who use their mobile devices; and 2) the podcaster that utilizes their laptop or desktop computer. I will lay out the hardware you want to get depending on which podcaster you want to become.

Although you can use your mobile device microphone to record your voice, in some conditions, you might need to get an external microphone. What I would recommend is that you test out your default microphone before you purchase any of these microphones. You may find that in a tranquil environment, your mobile device microphone is excellent. The ideal situation is conducting your podcast via laptop or desktop computer. You simply have more options for quality audio when you decide to use your laptop or desktop computer. Browse these industry-standard recommendations for your hardware equipment. Be mindful that the Android and Apple devices are without headphone 3.5mm jacks. Also, the new Apple MacBook laptops do not have USB ports anymore, so you will need to purchase a converter.

Product for Apple	Product for Apple + PC
RØDE - VideoMic ME-L	Blue Yeti Mic
Purchase from Best Buy or Amazon	Purchase from Best Buy or Amazon

Mac USB Converter
Purchase from
Best Buy or Amazon
https://amzn.to/3mzO7yY

Setting up your Accounts

There are four major components of this chapter and 1 of which you should have already handled. Setting up a separate Gmail account (should be already done), picking where you will distribute your podcasts, setting up accounts to organize guest appearances, and establishing your social media accounts.

Make separate podcast email

This item should have been completed in number 3.5 of the Executive Thinking:

Number 3.5 – Do not think about this one. Just do it now. Make a separate email account. Gmail by Google if you can. Gmail has a multitude of free products that will assist you perfectly with your podcast journey. You want to make sure all of your Podcast information is in one place.

Visit:
https://accounts.google.com/signup

After you create a free google account, you now have access to Gmail, Google Docs, Google Sheets, and Google Forms. These services from Google are formidable for keeping your podcast communication and coordination with guests organized. We will speak about how to use the different offerings from Google in a synergistic way later in the book.

Social Media Accounts

If you are just starting out, and you do not already have a large following of your own on social media, I recommend you create a Twitter and/or Facebook to represent your podcast.

I would recommend starting with one social media account first. To pick, I would do some research on both of the platforms to get a sense of what people are already talking about on the platform. For me, when I started PhD Tips, Twitter was buzzing! So I decided to start with a Twitter account. On both platforms, #HashTags are highly helpful to finding social media communities.

Where will your Podcast stream?

There is a multitude of streaming platforms you can get your podcast on. But what you need to decide is where you want your podcast to originate, which will distribute to a multitude of platforms. I will share the most accessible and easy-to-use platform, which is free for anyone to use.

Anchor

You may have seen Anchor on your news feed, or maybe you listen to other Podcasters talk about how they use Anchor. The major attractions of Anchor are the ease of use and automatic** distribution to popular streaming platforms like Spotify, Apple Podcast, and Google Podcast. I used ** because Anchor only automatically distributes after you connect your Anchor account to the various platforms. Anchor is a free app that enables creators of podcasts to organize their files, edit their segments, and publish their episodes to all podcast outlets people are listening on. So, if you've been saying, "I don't know how to start a podcast" because of technical know-how or motivational drive, the Anchor app helps you conquer both.

If you are ready to use Anchor, follow these directions. Let's go deep.

To get started, visit: https://anchor.fm/signup to create an Anchor account, using the new Gmail you created for your podcast. After you click sign up, make sure to confirm registration in the email Anchor sends you. You won't be able to publish any new episodes on your podcast without confirming.

Anchor will give you some recommended first steps, but here is what you should do.

1. Go to Settings >> Update Settings
2. Fill in your Podcast Name
3. Fill in your Podcast Description
4. If you developed your own Podcast Cover art, upload that.
    a. If you do not have your own cover art, you can pick some art that is provided by Anchor.

5. Select the appropriate Podcast Category and language of your podcast

Everything else in the settings is self-explanatory. Fill all of the information out so that it reflects your new podcast.

Setting Up Distribution

To get rolling with distribution, you must first release one episode. That episode does not have to be a full episode. I would simply make it the "Trailer" of your podcast that explains what your podcast is all about and what your goals are. This episode should literally be less than 2 minutes. The reason we need to release a trailer episode is that we need to get the RSS feed of your podcast to connect to other streaming platforms.

Recording First Episode (the trailer)

In Anchor, it is straightforward to post episodes. Click the big 'New Episode' button to proceed to the episode creation environment.

When you enter the New Episode screen, either on your phone or computer, there should be buttons that look like this.

Figure 14

As you learned in the Software section, you know that you can record some audio either in GarageBand, Audacity, or your mobile device. You can upload the exported audio file here to create your podcast. In addition, you are able to record an audio file right from the Anchor interface on the browser or on the mobile app. Whether you will record your files in the moment or do post-production in GarageBand or Audacity, make sure you upload the audio files here to set up your episode.

When your audio files are all uploaded, click the Save Episode button to be directed to the 'Episode Options.' In this section, you will be prompted to create your Episode title, description, and date + time you want to publish the episode. After this information is filled out, you will be asked to upload your Episode art, season number, and episode number. For the purposes of this trailer episode, make sure that the 'Trailer' episode type is selected.

That page should look like this:

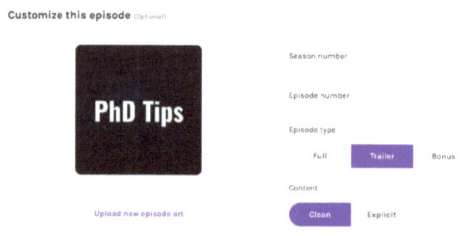

Figure 15

Once you have published your episode, the final step is to take note of your RSS feed.

To get your RSS feed, go to Settings >> Distribution. Your RSS feed should look like this:

Distribution

Your RSS feed

https://anchor.fm/s/14104910/podcast/rss `Copy`

Figure 16

Distributing Podcast

When you publish your Podcast on Anchor, the only platform it streams to automatically is Spotify and Anchor. What you will need to do for distribution on Apple, Amazon, Google, and all of the other big podcast streaming platforms, is place your RSS feed into their systems. The heaviest lift of this is to simply gather your RSS feed, which you already did. The only thing we have to do now is going to the correct links to connect your podcast. For all distributions, you should give the system about five days to connect successfully and eventually appear on your public Anchor podcast page. I will show you Apple, Google, and Amazon because they show three different processes that can be replicated to other podcasting distributions.

Apple Podcast

In order to get your podcast up on Apple Podcast, you must follow this link:

https://podcastsconnect.apple.com/

When you click this link, you will be instructed to enter your Apple ID. I recommend that you create a new Apple ID using the email you created for your podcast. After you have signed in, you should see a blue + button or an Add button that will prompt you to paste in your RSS feed. Paste it in, press submit, and you are finished! If you are running into any problems, follow this link to Apple Guidelines and troubleshooting issues:

https://help.apple.com/itc/podcasts_connect/#/itce5b9b0782

Google Podcast

In order to get your podcast up on Google Podcast, you must follow this link to Google's New Podcast Manager (make sure to use the Gmail account you created for your podcast):

https://play.google.com/music/podcasts/portal/u/0/start#

Submitting your podcast to Google Play Music is easy. Just follow these simple steps:

1) Click the blue "Add A Podcast" button near the top right:

ADD A PODCAST

2) Paste in your podcast's RSS feed URL
3) Click "Submit RSS Feed"

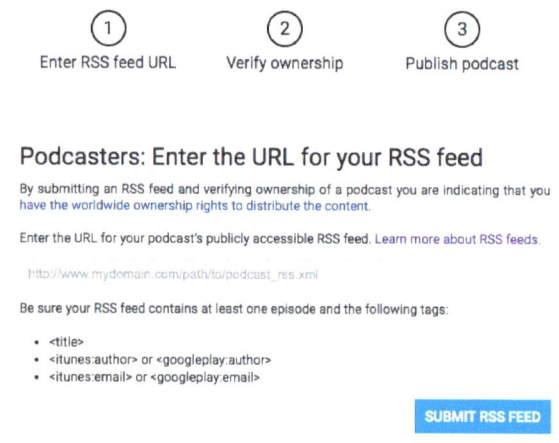

Figure 17

As you can see, you need to make sure you have at least one episode published, as well as these three tags:

<title>

<itunes:author>

<itunes:email>

If you have completed every step in this book so far, you should be all set with this.

1) Verify Ownership (uses email in RSS feed)

a) The email tag is where Google will send verification, so make sure you have access to it and that it is up to date.
2) Once verified, you're ready to review and publish your podcast. If any required info is missing from your RSS feed, you'll get an error with specific instructions on what to fix.
3) Click "Publish Podcast"
   a. If everything looks good, it's time to hit 'Publish Podcast'

Amazon Music

In order to get your podcast up on Amazon Music, you must follow this link to Amazon's Podcast Application:

https://www.amazon.com/music/lp/podcasts

When you scroll down a bit, you will see a link that looks like this:

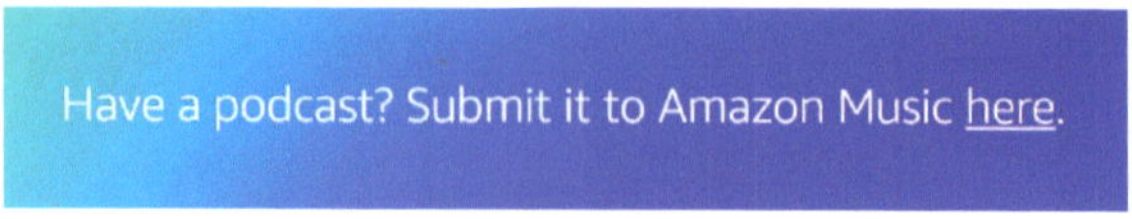

> Have a podcast? Submit it to Amazon Music here.

Figure 18

This is much simpler than Apple and Google. Amazon will walk you through a lengthy terms and conditions page, a section for you to insert your Podcast Name, and finally your RSS feed. All you need to do is fill out that information, confirm the details are correct, and then await for Amazon to approve your request.

Intake Forms and Templates

Since you have a Google account, you might as well utilize its power. Organizing information that you want to collect for your podcast can be done by utilizing Google Forms. The most common information you want to intake is for guest appearances and listener requests. As a gift from me to you, feel free to use the template intake forms in the links below for your podcast.

Guest Appearance Template Form:
https://forms.gle/Y7zmf51bT8p9gDr38

Listener Topic Request Form:
https://forms.gle/qXqyusUHJPxTwqRw6

How will you get the word out?

As we spoke about earlier in the book, there are many ways to get the word out about your new podcast. The three strategies I want to share with you are social media, email marketing, and paid advertisements. The best-case scenario is to use a little bit of all three, but if you find a sweet spot in 1, keep on crushing it!

Social Media

Ok, so by now, you should have figured out which social media platforms you want to use to represent your podcast. If you are a popular person already and the podcast is an extension of yourself, then your personal platforms will work perfectly, and you likely will not benefit from much in the social media section. But if you are not famous (yet), then continue on in this section.

Inner Circle

When launching your podcast, your first point of engagement should be your inner circle. Using Facebook as an example, your inner circle are your direct friends and family you are already connected with. For Instagram, Twitter, LinkedIn, or TikTok, that converts to your followers and/or connections. When making the first post to your inner circle, be you! Everyone that will see it knows you know what you're about and should be excited to know that you are starting a podcast. So share with them WHAT and WHY you are starting your podcast, incorporating as much of YOU in the post as possible. If you can, try to make your post on a monumental date, like your birthday or a popular holiday. When making your announcement to your inner circle, you want it to be an invitation, not an obligation (forced engagement). More on Forced Engagement later. On the eve of the 4th of July (United States Independence Day), here is the post that I made when I launched PhD Tips:

I'm excited to announce that I've signed myself up to running my very own Podcast named PhD Tips!

So here is the idea. As I plan to stumble and fumble through my own PhD program, I will record podcasts to share key learning points, milestones and general tips for others who may land in my shoes one day. Along the way, I will invite current PhD students, Post Docs, and PhDs in the field to share some of their very own insights and tips on "how to get through your PhD Program successfully."

It feels weird calling this podcast "my very own" because I'm hoping the exchange of information, ideas, and tips discussed on the podcast become a sort of north star for the culture (aspiring Black PhDs). Because in my very straight forward opinion...WE are our best shot at being great. Many of the voices you will hear on the podcast are going to be of some of the most brilliant black minds in their space.
So enjoy and listen to PhD Tips on your favorite podcast streaming platform.

And while I have your attention...on the eve of the 4th of July... go read Fredrick Douglass's "What to The Slave is The Fourth of July?

Forced Engagement

Forced engagement is basically spam mail. You know those people who post the same thing every day, demanding that people support them? It's like... hey bro; you do know you keep posting to the same circle of people, right? If your following does not grow, then you are literally sending people the same message every day, which begins to make people feel "obligated" to view your stuff or completely annoyed with you... because everyone hates spam. We will learn a little about how to grow your circle in the advertisement section. The moral of the story about Forced Engagement is, you do not want to continuously tell people to engage with your stuff... because if it's nice stuff, they will come. You do not want to appear as a charity case. And you won't as long as your posts are inviting and you are offering something people want/need.

Organic Traffic

As I spoke about earlier in the book, #HashTags and Search Engine Optimization are extremely powerful, especially for the purposes of organic traffic. Organic traffic is simple to understand but takes good patience and keyword optimization. As we spoke about in the Search Engine Optimization and Google Trends section, keywords can make or break your organic traffic attempting to reach. One audience, and the primary one, is the prospective Ph.D. student or current Ph.D. student who wants to flourish in their program. The second audience is a person who can authentically provide tips to someone on their journey to become a doctor. Conveniently, social media platforms have coalesced around the idea to allow users to create communities within their platforms. This option makes it really easy to find groups that are already formed with the set of characteristics you are looking for. In my case, if I ever need some guest speakers or featured blog post authors to submit some written tips, I make a post in a group of recently graduated Ph.D. students. When I feel like my listenership is plateauing, I make periodic New Podcast posts in groups where there are recent graduates of Grad School or Career Inspiration groups.

Email Marketing

Email Marketing's major goal is to market your content to end-users so that they can engage with it. Email Marketing is tricky first starting out because you have not established an email list yet! For starting out, a good strategy to start collecting emails is FREE giveaway opportunities. This works well when you have some kind of free service you can give, online store discount codes, or other monetary items to give away for free in exchange for an email. As a plus, when the opportunity you are offering is really good, people will share the opportunity with their own inner circles. When that happens, you are golden! What you want to do is set up a system where people will have to either email you or express their interest in your opportunity. The easiest way to do this is to create a Google Form where you ask for your name and email. At the end of the timeframe you set for the opportunity, all you would need to do is pick a random person from the list and send them their reward. Make sure to let your listeners know who won on a future Podcast so that they know it was real. Once your giveaway is over, you now will have a number of emails in your Email List, which you can use to send updates to every time you have a new Podcast. Make sure to monitor your listenership after your giveaways to see if your streams increased. If it did, then that means you successfully increased your listenership, and the giveaway goal of increasing your streams was successful.

Closing

If you have made it to this point of the book and have read every last word, you are fully prepared to start your podcast. Nothing else is in your way besides you finding time within your day to dedicate a little bit of time to chip away at all of the content in this book. When I started my podcast, it took me roughly three weeks or so to get it fully rolling and out to the public to listen. Over time you will continuously find ways to optimize your workflow and pivot the themes of your podcast as you and your listeners need. Most of my time now as a Podcast owner is devoted to finding avenues for where I can place the podcast so that more people can listen. You will get to a point where you will need to expand your Podcast reach by developing a Podcast website, maybe you will start selling merch, become a paid sponsor, or even you will get so big that people will pay you to be a guest on your show! Your eagerness to be great is the only limit, so believe in yourself and remain consistent. I promise you; this will all be rewarding in the end.

About The Author

Cliff Freeman was born and raised in Dorchester, a resilient neighborhood in the city of Boston, Massachusetts. As a young boy growing up, Cliff was always fascinated with technology and how things worked. In high school, he was the kid who roamed the hallways with a camera around his neck and handled all of the technological breakdowns that came up at school. In 2010, at the age of 16, Cliff became wildly interested in learning more about technology when he made his first "big ticket" purchase: A 2010 Dell Inspiron Laptop. He taught himself everything about his new laptop by deconstructing it, replacing parts for better parts, downloading software, tinkering, and even teaching himself the essence of HTML programming. Cliff's love for technology has guided his journey to where he is today, passionate about the potential of technology and the opportunities it can open for marginalized peoples globally. Cliff's work is currently at the intersection of social justice, science, technology, engineering, computer science, mathematics and education research. The category of books Cliff mainly writes are technological in nature that guide people through on how to use technology as a tool to thrive in the 21st century.

Learn more about cliff @
https://cliffslink.com/

Works Cited

Google Search Statistics—Internet Live Stats. (2020).

Internet Live Stats.

https://www.internetlivestats.com/google-search-statistics/

Google Trends. (2020). Google Trends.

https://trends.google.com/trends/?geo=US

Statista. (2020). *Facebook: Daily active users worldwide*.

Statista.

https://www.statista.com/statistics/346167/facebook-global-dau/

www.ingramcontent.com/pod-product-compliance
Lightning Source LLC
Chambersburg PA
CBHW040324220526
45473CB00009B/2560